Beyond The Salt Shaker - How Salt Changed Mankind

By:

Kellie Graham

Copyright © 2016 Kellie Graham

Smashwords Edition, License Notes

This ebook is licensed for your personal enjoyment only. This ebook may not be re-sold or given away to other people. If you would like to share this book with another person, please purchase an additional copy for each recipient. If you're reading this book and did not purchase it, or it was not purchased for your enjoyment only, then please return to your favorite retailer and purchase your own copy. Thank you for respecting the hard work of this author.

Table of Contents

Where Does Sea Salt Come From?

When Did Humans First Discover Sea Salt?

How is Sea Salt Formed?

How is Salt Extracted from the Sea?

What is Unrefined Salt Used for?

What Makes a High Quality Salt Variety?

Why is Salt So Important for Human Societies?

How Was Salt Traded Across the World?

What Happened During the Salt Wars?

Why Does Salt Appear in Many Holy Texts?

Is Salt Still As Important Today?

How Did Salt Preservation Change Society?

What Are Some of the Key Dates in the History of Salt?

The Present Day

The Benefits of Unrefined Sea Salt Products

Throughout the past decade, salt has been a topic of great debate across the health and diet industries. People are routinely told, 'Cut back on salt intake.' and 'Avoid foods with too much added salt.' but this is not always as simple as it sounds. Much of the prepackaged food sold in supermarkets is packed with sodium. Western populations have come to be dependent on it as the normal taste; this has happened in the same way that they have slowly become addicted to sugar. Did you know there is sugar in iodized salt? Take a look on your container, it lists dextrose. The kind of salt you consume matters.

Yet, the argument is much more complex than one might expect. While there may be compelling evidence to support the destructive nature of concentrated chemical salts (like table salt), there is no such evidence to suggest that unrefined and unprocessed salt is similarly harmful. In fact, scientists and health experts are now beginning to recommend unprocessed sea salt, as it proves to be beneficial for the body and diet.

Read on and we will explore the ways that sea salt has influenced human culture, development, and diet. We will discuss some of the key advantages of its use and explain where it comes from, how people get on their plate, and why it deserves a little more acknowledgement as a health food.

Where Does Sea Salt Come From?

We consume salt daily, but have you taken a moment to think about where salt comes from and how it is collected? Most people are well aware that salt originates from the sea. Did you know that salt can also be mined, directly from the ground, in the same way that coal or diamond is?

Salt's ubiquitous taste is recognized by everyone in the world. The human tongue actually has a taste bud dedicated to salt. It is most abundant within the seas on our planet Earth, but many ancient bodies of water still exist in the form of dried up salt deposits. There is an average of 26 million tons per cubic kilometer of salt in sea water. If extracted, the total amount would coat the surface of the planet in salt, up to a depth of 35 meters. So, it is fair to say that human beings aren't running out of salt anytime soon!

There are endless uses for salt. Salt is added not just to food, but also to things like roads and highways to prevent them from freezing. In some areas of the world, salt is even used as a substitute for bullets, so that self-defense means incapacitating and not fatally wounding the target. With all of these diverse uses, it makes sense that salt comes in a wide variety of different forms.

What do you know about unrefined sea salt? This refers to salt which has been collected from the ocean using natural and non-invasive methods. 'Natural' in this context, means that the salt has been extracted from the water via evaporation. No chemicals or artificial compounds have been added to stimulate or speed up the process.

According to health experts, this is the right type of salt for humans to be consuming. While it is always best to eat salt in moderation there is a big difference between eating salt that contains added sugar and chemicals and salt that has been 'lifted' from the water. Changing out the salt you use can be a really vital component when trying to maintain health or improve diet.

When Did Humans First Discover Sea Salt?

There has never been a time, certainly that we know of, which has seen salt take a back seat for human beings. It has always been of huge significance. From the dawn of the earliest civilizations, people have collected salt from the sea and used it for food production and trade. The earliest known town in Europe, Solnitsata, was built around a salt production site.

It was located in what is now Bulgaria and historians believe that the town sustained its wealth by trading salt right across the Balkans. In the United Kingdom, the suffix '-wich' was added to towns which took on this role. Regions like Sandwich and Norwich were big producers and traders of sea salt. Its significance should certainly not be underestimated, because most archaeologists recognize that it shaped the course of human history.

Not only did it open up new possibilities for food production, it was also used in war as well. 'Salting the earth' involves spreading salt across large patches of valuable ground, in order to kill the essential nutrients in the soil and make it hostile. This military technique is very old and was first practiced by the Assyrians.

It is easy to see how important salt has been for human culture when words like 'salary' are considered. The term is derived from the Latin word for salt, because Roman soldiers were often paid their wages in salt. As the substance was extremely valuable, skilled soldiers were said to be (quite literally) 'worth their salt.' In the modern day, however, salt is extremely affordable now that transportation is more efficient. This doesn't make it any less important though, so human beings have come up with a number of different ways to produce and use it.

How is Sea Salt Formed?

In a unrefined, unprocessed state, sea salt is mostly made up of sodium and chloride. It also contains trace elements, which is why, in healthy amounts; unrefined salt can be a valuable dietary addition. Unprocessed salt is the salt directly from the ocean and available commercially. These are the pure salts from the oceans!

Recently, there has been a huge rise in the number of unrefined salt products entering the market. Today, there is a very lucrative industry built around the appreciation for unprocessed salt. It is not just better for the body, but also much better tasting too. Reliable vendors, like Salts Worldwide, sell a broad range of salt products, all sourced from specific regions of the planet Earth.

Different regions produce different flavors. For instance, an unrefined salt product from a Hawaiian sea won't necessarily taste the same as salt from around the Baltic regions. According to region, unprocessed salt may have slightly different compounds and proportions of minerals.

A good example of this is Himalayan Pink Salt, which contains more than 84 trace elements and minerals.

[Pink Himalayan Salt](#) is just one example of an unrefined salt product which is highly sought after for its health benefits. Its calcium, magnesium, potassium, copper, and iron compounds offer the body a multitude of fortifying minerals. According to health experts, consuming Himalayan Pink Salt on a regular basis balances electrolytes, supports efficient nutrient absorption, eliminates toxins, and stabilizes PH levels.

It is also thought to normalize blood pressure, enhance blood flow, and alleviate the pain associated with conditions like arthritis, skin inflammation, herpes, and fever and flu symptoms. Himalayan Pink Salt is not the only choice, because there are many effective, unprocessed salt products on the market. To check out their benefits, one only needs to swap out regular, processed table salt for an exotic, raw salt variety.

How is Salt Extracted from the Sea?

'Solar evaporation' is the term given to the most natural and unrefined process of extraction. It is how human civilizations have been harvesting salt for centuries. To feel the full benefits of unprocessed salt products, the substance must have been collected in this way. The process is extremely simple and it has remained largely unchanged for many hundreds of years.

First, salt water from the sea is directed into shallow pools. Here, it is exposed to the hot sun and wind. Together, the heat and exposure to the elements causes the salt to be pushed out of the sea water. It evaporates and a pile of unprocessed salt develops on the bottom of these little pools. The salt is collected, washed, quality checked, and shipped out all across the world.

Currently, the average solar 'crop' can take anywhere from one and five years to 'rear' and complete. It is important to understand that this technique is the only one associated with unrefined sea salt. Unrefined products should be as close as possible to their pre-evaporated condition. No chemical agents should be used to extract or alter the molecules of the salt.

Doctors and health experts generally agree that unrefined salt is much better for the body than processed varieties. However, it should still be consumed in moderation. The Dietary Guidelines for sodium, in the USA, recommend no more than 2,300 milligrams per day. This figure is lowered to 1,500 milligrams for people over the age of fifty and those with pre-existing health conditions like diabetes.

What is Unrefined Salt Used for?

As aforementioned, salt can be used for a wide range of different things. Primarily though, human beings tend to use it for flavoring on food. Historically, it was the main method of food preservation. Without salt, much of the trade carried out, all around the world could never have developed in the first place. These days, salt is still used to preserve ingredients like fish and vegetables, but it is employed for flavor with each meal.

The soaring popularity of unrefined and unprocessed salt varieties has seen millions of health conscious consumers turn to healthier products for home cooking. While unrefined salt crystals are larger, tougher, and less polished than standard table salt varieties, this is actually no bad thing.

It is becoming increasingly common for people to use unprocessed salts as a beauty product. They are added to bubble baths to soften the skin, mixed with honey to treat inflammation, and combined with olive oil to create gentle, all natural sea salt scrubs. There is also a small amount of evidence to suggest that sea salt inhalers can alleviate the symptoms of asthma and allergy. This is very new science though and it should be considered with the appropriate degree of caution.

Unrefined sea salt is thought to have anti-aging properties, antibacterial qualities, and it may even help to re-mineralize and strengthen dental enamel. Its uses are varied, diverse, and very effective, so it is worth checking out to offer unprocessed salt varieties in your kitchen. Adding just half a cupful of salt to a hot bubble bath is a great way to soothe the skin and reinvigorate tired muscles.

What Makes a High Quality Salt Variety?

When shopping for unrefined salt products, read labels to make sure that you are buying exactly what you think you are. Usually, the manufacturing process is clearly outlined on the packaging. There should be no added anti-caking agents in unrefined salt varieties, because they are designed to resemble, as far as possible, the salt in the oceans and seas.

Do be aware that unrefined salt products can come in a wide range of different colors, shapes, and sizes. This variance is a good indication of its purity and 'all natural' characteristics, because nature creates in all kinds of colors.

While unrefined salt varieties almost always seem more costly than table salt products, it is important to remember that a little goes a long way. Authentic sea salt is stronger in the salt flavor, you can use less and taste more. The salts are much better for the body. Either way, sodium consumption should be carefully controlled, but unprocessed salt makes this easier, because it adds essential minerals and nutrients to a diet. Plus, it tastes significantly better than standard table salt varieties. It would never contain added sugar dextrose.

It can be enjoyed with food in exactly the same way as a person would use table salt. However, some crystals are larger, but there is also a finer grain available, it is probably best to add smaller amounts or cook the ingredient into meals before eating. The finer grain is best when used in cooking for perfection of measuring. Whether unrefined salt is sprinkled over pasta, potatoes, vegetables, stir fries, or greens, it is guaranteed to enhance quality, flavor, and satisfaction. With a high quality unrefined salt variety, all of the natural benefits of the sea are combined in a wholesome, unique flavor.

It is possible to create salt artificially, but it exists in such abundance that there is no real need for this kind of science right now. Interestingly though, the US salt market does not discriminate between natural and artificial salt varieties, as long as both meet stringent purity requirements. In other words, the salt that is being sold to consumers does not have to have been taken from oceans and seas.

Health experts do agree that natural products are significantly kinder to the body. They boost levels of essential minerals and eliminate the need for nasty chemicals and complex manufacturing techniques. The question is, how did the salt industry reach its current position? Just how big of a role has salt played within the development of human history?

Why is Salt So Important for Human Societies?

When exploring the history of salt, the first place to start is always with preservation. Yes, flavor and seasoning has long been coveted, but the ability to preserve ingredients changed the course of civilization. This is no overstatement either. With the power to preserve comes choice. For a hunter gatherer, living in a dangerous world, food preservation meant the difference between risking life and limb for food every day or a couple of times per week.

This freed up so much time for further exploration. People were not tied down to the inevitability of the next hunt, because they could save their food for later. The time that would usually be spent hunting or gathering could, instead, be used for drawing, singing, poetry, building, trading, and farming. So, food preservation really did change the face of humanity and salt was a key part of this.

With the ability to preserve, also came the power to trade over longer and longer distances. Historians know that salt was one of the first commodities to be traded extensively, all across the world. This ended up having a profound influence even on the way that countries operate today. For instance, during the early years of the Roman Republic and the expansion of Rome, dedicated roads were created to aid the transportation of salt in and out of the city.

One of the most well-known examples of this is the Via Salaria, a road leading outwards from Rome to the Adriatic Sea. This was a popular transportation and trading route, because the Adriatic waters are shallower and filled with more salt than all of the other seas in close proximity to the city. At this time, salt was a very valuable trading commodity, because it could only be sourced from the sea.

The sprawling salt deposits under the earth were initially out of reach, so unrefined salt became something that every society wanted. Its value can be observed in common phrases like 'worth your salt,' which was used to describe skilled Roman soldiers. Salt was almost as valuable as gold, it was not unusual for soldiers to be paid their wages in this alternative currency. Hence the term 'a person who is worth their salt.'

It is not just financial value that is relevant here either, because the phrase 'salt of the earth' continues to be used even today. This is a biblical term which means good and honest. If a person is described as 'the salt of the earth,' they are being emphatically complimented. It is just one of many examples of salt being infused with value, meaning, reliability, and gratitude. As the long and illustrious history of the human relationship with salt shows, dependence on nature and its naturally occurring commodities is essential for survival.

How Was Salt Traded Across the World?

Historically, the international salt routes were nothing short of remarkable. They spanned the whole planet, from Greece to Timbuktu. One of the busiest routes stretched from Morocco and across the Sahara desert. However, ships loaded with salt also regularly made their way across the Mediterranean and the Aegean.

The ancient Greek historian Herodotus talks of a sprawling network of 'caravan and trade routes' which united the salt oases of the Libyan sands. It has become custom to think of early cities like Venice as being built on the top of strange and exotic spices, but the reality is that humble salt offered the biggest moneymaking opportunities.

During the early 6th century, in sub-Saharan Africa, Moorish traders regularly exchanged salt, ounce per ounce, for gold. In Abyssinia, salt was harvested and packed down into the shape of rudimentary 'rock coins.' These bricks of salt were taken on long voyages and used as currency in various parts of the continent. In all of its infinite abundance and availability, salt still managed to become one of the most important (if not the most important) commodity throughout the history of human development.

It wasn't just the ability to preserve and flavor food that salt trading opened up either. For many centuries, unrefined salt has been used as a form of antiseptic. Even today, dentists advise patients to rinse their mouths with salt water after a treatment or procedure. It helps to close up the wound, eliminate nasty bacteria, and form a protective barrier. The Romans were well aware of this, because their word for salt ('sal') is derived from the name for the goddess of heath, Salus.

There were many roads that a trader could take to reach Rome, but by far the most popular was the aptly named Via Salaria (or 'the salt route'). Every year, thousands of salt deliveries would make their way along this route and end up in the hands of savvy city shopkeepers. Throughout history and the rise and fall of civilizations, this pattern continued. Over time, the value and significance of salt would increase even further.

What Happened During the Salt Wars?

The role of salt has been surprisingly prominent in the fate of major cities. Wars have been started over salt. It allowed industrial towns like Liverpool, in the UK, to progress from tiny ports to sprawling hubs of trade and commerce. During the 19th century, Liverpool produced a huge proportion of the salt consumed and exchanged across the entire world.

Salt has nourished and destroyed empires. In the 16th century, Poland grew rapidly, after it discovered a wealth of underground salt mines. The country became very wealthy, very quickly, until Germany accelerated its production and trade of natural sea salt. Health conscience people make a choice to use natural sea salt over processed rock salt. It was thought to taste better, add more flavor, and compliment the body.

The passage of salt gave smaller countries the opportunity to grab a little influence and power. Many cities, duchies, and states positioned along key salt routes demanded hefty tolls for safe passage through their territories. In 1540, the desire for salt caused major trouble in Italy. Pope Paul III, in a bid to take back the semi-autonomous rule given to the city of Perugia, exacted a heavy tax on salt.

This caused huge problems for the citizens of the city, who struggled to pay the price and continue preserving their food. Eventually, the people of Perugia decided to rebel and they attacked the papal territory. Their uprising was quickly quelled and an enormous fortress was built to put distance between the Pope and the city. For many centuries, this oppressive fortress was a dark symbol of the power of Papal rule.

Interestingly, an urban legend still persists about the infamous Salt Wars. It is thought (though never confirmed) that the traditional breads customary to Perugia (even today) are unsalted because the rebellious citizens boycotted the use of the ingredient. In India too, during the thirties, salt became a symbol of resistance and independence.

In 1930, celebrated activist and independence leader, Mahatma Ghandi, led Indian citizens to the sea to collect salt. This action was initiated after the occupying British forces tried to physically prevent people from manufacturing salt in this way, as they had done for many centuries. In response to violent enforcement measures and the idea that Indian people should have to pay the British for salt or be branded criminals, Ghandi and his followers publically defied the ban.

They marched for twenty four days to reach the coast, in Gujarat, and they were joined by huge numbers of supporters along the way. The protest was peaceful, but it provoked many instances of violent rebellion and reactionary oppression between the British and the Indians. This was a turbulent period in the history of India and while the tax on salt was not the root cause of the tension, it did play a part in aggravating it.

Why Does Salt Appear in Many Holy Texts?

It doesn't take long when browsing holy texts to spot mention of salt. This ingredient has long held an important place in religion and culture. For Hindus, it is considered to be a very lucky and positive substance. The traditional salt ceremony of 'Datar' sees new brides pick up a handful of salt and hand it to their husbands. They repeat this three times, while being extremely careful not to drop any grains.

The ritual is carried out in the presence of both families and it symbolizes their union, as a result of the marriage. Just like salt in a tasty meal, the bride and groom (and their loved ones) have become permanently blended together. Alternatively, many Buddhists use salt to ward off evil, particularly after funerals and personal crises. Similarly, Shinto worshippers use salt to purify people and locations.

They do this by leaving small piles of salt in little dishes by the entrances of buildings that they believe need purifying. As for the Bible, it contains one of the most infamous stories to feature salt that has ever been told. In Genesis 19:26, Lot's wife is instructed not to turn back and witness the destruction of the lost cities of Sodom and Gomorrah. She disobeys this order and is turned into a pillar of salt as punishment.

There are plenty of different interpretations of this event. It is an ambiguous one for lots of reasons. For one thing, Lot's wife was mostly innocent. Yes, she disobeyed the order not to look back, but it is generally assumed that she was searching for her daughters and evidence of their escape. From this perspective, the punishment is an extremely harsh one.

The story is further complicated by the fact that salt was generally considered to be a very valuable commodity during Biblical times. It is used, again and again, in the holy book, to signify permanence, loyalty, durability, and fidelity. So, the question of why God chose to turn Lot's wife into a pillar of salt continues to be hotly debated. There are some theologians who believe that she was sacrificed, so that the destroyed cities could have a guardian.

Is Salt Still As Important Today?

Either way, the significance and influence of salt, throughout history and the development of human civilization, cannot be denied. This humble ingredient, which exists in abundance across the planet, has been present at the birth and fall of nations. It has toppled kings and allowed underdogs to seize their opportunity to make a mark on the world.

Today, it has less obvious influence but it remains ever present in all societies. Without salt, people would lose the power to quickly and easily season ingredients. The most effective and affordable form of preservation would become inaccessible. Life would surely be a lot blander. Yet, would people actually be healthier? Is it that salt still essential for good physical health; or should modern diets include much lower amounts of sodium to begin with.

The answer depends not only on how much salt a person consumes, but also what kind of salt. Unrefined and unprocessed varieties are much kinder to the body than standard table salt products, because the latter is treated with agents and additives. With unrefined sea salt, the aim is to get the ingredient to the dinner table or the kitchen cupboard with as little intervention and change as possible.

The value of salt, throughout history, is one of its most remarkable features. It was once, quite literally, worth its weight in gold. Roman soldiers were routinely paid in salt, because it was as good as giving them coins. In some ways it was even better, because salt was harder to access than gold. It also meant that the soldiers could take their wages (or 'sal-aries') home and use them to keep their families fed.

The ability to preserve food was one of the reasons why salt was so valuable in ancient times. While people living in colder parts of Europe could use icehouses and ice stores to freeze their meat and other ingredients, most of the continent didn't have this option. The sun was too hot and the temperatures too high. Access to salt meant that families didn't have to constantly worry about sourcing food.

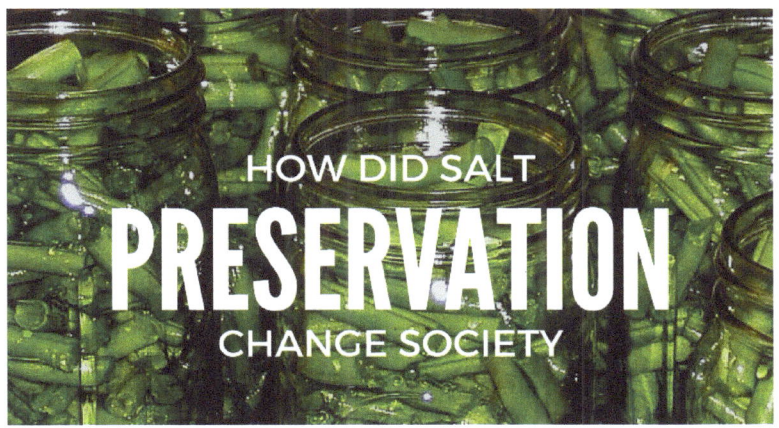

How Did Salt Preservation Change Society?

The ability to preserve and store food has benefits. It significantly reduces the risk of a family starving, as ingredients can be rationed. Without salt, Europeans had to be content with eating as and when the opportunity arose. So, there were certainly immediate benefits. The question is, how did salt preservation change human societies?

Well, it all starts with choice. The power to decide whether or not to eat or store food is something that not many species have. Lions, bears, wolves, and other carnivores must hunt their food, eat it quickly, and then find more. This means that their lives revolve constantly around hunting. When they are not hunting, they are eating. When they are not eating, they are searching for something to eat.

So, choice is very important. With access to salt and salt preservation techniques, came the freedom to focus on other things. Human development is pushed forward not just by fundamental needs, but also by leisure; when people are not constantly striving to survive, they can write, draw, chat, dance, sing, and pursue all kinds of non-vital activities.

Ultimately, salt influenced the development of human societies, by giving them the freedom and the time to evolve. From this perspective, it suddenly doesn't seem like such a stretch to realize that sea salt used to be as valuable as gold. It might not be quite as expensive today, but it still offers many different benefits to the diet and health.

For instance, unrefined and unprocessed sea salt is now commonly used as a skin softener. It is added to baths as a way to gently remove dead skin and stimulate the production of new cells. It can also be used to treat inflammation, alleviate the symptoms of certain allergies, and eliminate bacteria in the mouth and around minor wounds. Today, piercing studios still recommend that customers wash their new jewelry with salt water.

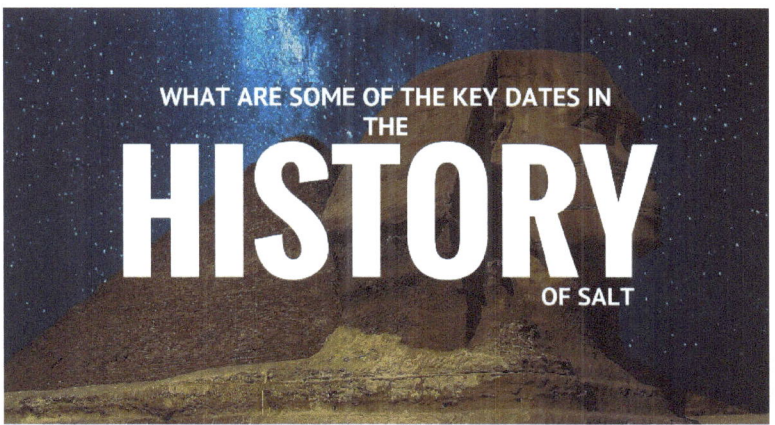

What Are Some of the Key Dates in the History of Salt?

This is a brief guide to some of the most important dates in the history and development of salt. There is a neat infographic that explains the history of salt.

10,000 BC

At this time, human beings were just starting to discover the salt in the oceans. They were also beginning to realize the benefits. Until this point, communities existed on grains like rice, barley, wheat, and millet. They ate very little meat because it did not last for long. It could not be preserved, so despite being very nutritious, it was highly impractical.

Once salt started to be used for preservation, all of this began to change. Farmers, fisherman, and hunters took the opportunity to produce and store things like salted cod, bacon, and ham. They did not have access to refrigeration, but were able to preserve huge amounts of fish and pork in big barrel of salt from the sea.

6,000 BC

In China, at this time, one of the earliest systematic salt works ever discovered was created. It extracted salt from the surface of the Xiechi Lake. This marks the beginning of the salt industry, because it allowed the rate of production to significantly increase.

Also, the Assyrians became the first people to start salting the soil in enemy territories. They did because they knew it would ruin the health of the earth. If salted heavily enough, farmland will never recover and communities must relocate or starve.

3,000 BC

In Egypt, during 3,000 BC, communities were trading salt as a commodity. The Egyptians routinely exchanged salt with the neighboring Phoenicians, in exchange for a range of different luxuries and treasures. This small scale trading gave birth to something much bigger, because it represents the humble beginnings of the international salt industry.

The Phoenicians took the salt from Egypt and used it to trade with other communities, across the Mediterranean and North African territories. As the trade routes widened and grew, fortifications were put in place to protect them. The Saharan salt routes were heavily guarded and the communities that lived alongside them could sometimes charge for passage.

100 BC

Salt production has become a vital part of the Roman Empire. The bustling Via Salara (Salt Road), which stretches from Rome to the Adriatic Sea, is constantly filled with traders and dealers. The Roman Republic exerts a lot of control over the market price of salt. This humble ingredient becomes an effective way to fund wars and military campaigns.

Around this time, the word 'salad' was first invented. Like the word 'salary,' it too is derived from the Roman word for salt (sal). In Roman times, a salad was a meal of salted leaf vegetables. Over time, the scale and speed of salt trade continues to expand and, soon, forty thousand strong caravans are being used to deliver huge quantities.

1000 AD

As the salt industry develops, more and more city states and duchies attempt to profit from it, even if they have to do so indirectly. Many of these communities start requesting heavy taxes for safe passage through their territories.

This caused turbulence and tension across much of Europe. However, it also allowed some smaller states and communities to grow very rapidly. Over the years, access to salt only grew in importance and kings, bishops, and leaders were all out to get their hands on it.

1500s

During this time, one of the most infamous 'salt wars' occurred. In 1540, in Italy, Pope Paul III tried to gain control back from the semi-autonomous city of Perugia. He did this by putting a heavy tax on salt in place. The citizens of Perugia, who had always enjoyed easy access to salt and salt preservation, found it very difficult to feed their families.

Many could not afford the high tax and the decision was made to mobilize and attack the Papal territory. Unfortunately, the people of Perugia were unsuccessful and they had to surrender. The Pope then built a huge, oppressive fortress around the Papal territory and the relationship between it and Perugia grew increasingly sour.

1600s

During the 1600s, salt production played a vital part in the development of early America. It was used as a weapon, by both the colonies and the British occupiers. The Massachusetts Bay Colony began to product salt, but the British responded by cutting off their supply. This was a heavy blow because it prevented the rebels from storing and preserving food.

1800s

During the 1812 conflict, salt was used as a currency. It was given to soldiers in the field, in place of cash, because the government did not have enough money to pay them in the traditional manner. Around this time, a huge salt supply is discovered close to the Missouri River.

1930s

In India, during the 1930s, the British occupiers place a hefty tax on salt. In order to make sure that this tax is paid, they also introduce harsh penalties for anybody found collecting salt from the sea. The new rules were enforced with violence, on many occasions, and the response from the Indian people was rebellion.

Mahatma Ghandi led a group of 10,000 people on a twenty four day march to the sea. Once there, they staged a peaceful protest by defying the ban and continuing to collect their own salt. While the march was met with more violence and turbulence, it marked an important milestone in the fight for independence across the country.

1994

In Great Britain, the first public health guidelines on sodium intake are released. For the first time, health experts advise people to limit their salt intake to 6 grams per day. This is after evidence is found to support the detrimental impact of high sodium consumption on blood pressure and the heart.

A few years after these guidelines were released; scientists provided clear proof of a link between high sodium intake and high blood pressure. From this point on, sodium is treated very cautiously by public health authorities.

2005

At this time, the top five producers of salt, across the planet, are America, China, Germany, India, and Canada. Salt has started to lose its extremely high value, but it is still an important commodity. The difference between now and a hundred years ago is that salt is so easy to access.

Production processes are fast, efficient, and reliable. Global salt production is estimated to stand at around 210 million metric tons, every year. There is a burgeoning industry devoted to unrefined and unprocessed sea salt products. This is because health experts are starting to discover the difference and many benefits natural sat varieties.

The Present Day

Today, salt remains valuable, but it carries a stigma of unhealthy. Almost every country on the planet enjoys easy access to salt, so it can no longer be sold equivalent to gold. Yet, this does not stop it from being an essential part of modern life. For most populations, it is a vital component of cooking and eating.

While it may not be used quite as often for preservation any more, it is a staple on dinner tables, restaurant counters, and kitchens large and small. More is known about the health benefits (and downsides) of sodium consumption, people can make better choices about how to use it.

For instance, take unrefined sea salt products as an example. Over the centuries, salt production processes began to rely on more and more chemicals, as a way to speed up manufacturing and get deliveries out faster. Now, the need is less urgent and consumers are moving back to more natural and unrefined varieties.

The Benefits of Unrefined Sea Salt Products

Unrefined sea salt is salt that has not been chemically altered. It is taken from the sea and extracted via the use of evaporation and hand harvesting techniques. This is as natural as it gets. Without additives or sugar, the ingredient is as healthy and kind to the body as it can be. Unprocessed salt varieties, like the ones available from Salts Worldwide, are hugely versatile.

They can be used to flavor food, in the traditional way, or they can be used for a range of alternative purposes, such as cleaning and bathing. For instance, salt extracted from the Dead Sea is believed to have remarkable benefits for the health of the skin and complexion. Unrefined salt products are often used as part of skin and hair masques, because they provide a gentle exfoliation action, without being too abrasive.

Reputable vendors and retailers, like Salts Worldwide, who dedicate themselves to bringing the power of natural sea salt into homes, are the best place to start when it comes to exploring real, raw salt. At the moment, Himalayan Pink Salt is one of the most popular choices, but it is chased closely by Hawaiian Black Lava Salt and Fleur de Sel Sea Salt.

The best advice for buying high quality unrefined sea salt products is to opt for a company that is reliable, informative, and able to produce good reviews and testimonials from previous customers. All products should clearly state where the salt was sourced from and how it was extracted. There should be no agents involved in this process. Pick a reliable supplier and it is possible to enjoy unlimited access to some of the most nourishing sea salts in the world. After all, we only exist because of the salt content in our mother's womb!

Source - https://saltsworldwide.com/blog/the-history-of-salt/

www.ingramcontent.com/pod-product-compliance
Lightning Source LLC
Chambersburg PA
CBHW041147180526
45159CB00002BB/750